Highlights

Highlights of GAO-12-442, a report to congressional requesters.

April 2012

DEFENSE BIOMETRICS

Additional Training for Leaders and More Timely Transmission of Data Could Enhance the Use of Biometrics in Afghanistan

Why GAO Did This Study

The collection of biometrics data, including fingerprints and iris patterns, enables U.S. counterinsurgency operations to identify enemy combatants and link individuals to events such as improvised explosive device detonations. GAO was asked to examine the extent to which (1) DOD's biometrics training supports warfighter use of biometrics, (2) DOD is effectively collecting and transmitting biometrics data, and (3) DOD has developed a process to collect and disseminate biometrics lessons learned. To address these objectives, GAO focused on the Army and to a lesser extent on the Marine Corps and U.S. Special Operations Command, since the Army collected about 86 percent of the biometrics enrollments in Afghanistan. GAO visited training sites in the United States, observed biometrics collection and transmission operations at locations in Afghanistan, reviewed relevant policies and guidance, and interviewed knowledgeable officials.

What GAO Recommends

GAO recommends that DOD take several actions to: expand leadership training to improve employment of biometrics collection, help ensure the completeness and accuracy of transmitted biometrics data, determine the viability and cost-effectiveness of reducing transmission times, and assess the merits of disseminating biometrics lessons learned across DOD for the purposes of informing relevant policies and practices. GAO requested comments from DOD on the draft report, but none were provided.

View GAO-12-442. For more information, contact Brian Lepore, 202-512-4523, leporeb@gao.gov.

What GAO Found

The Department of Defense (DOD) has trained thousands of personnel on the use of biometrics since 2004, but biometrics training for leaders does not provide detailed instructions on how to effectively use and manage biometrics collection tools. The Office of the Secretary of Defense, the military services, and U.S. Central Command each has emphasized in key documents the importance of training. Additionally, the Army, Marine Corps, and U.S. Special Operations Command have trained personnel prior to deployment to Afghanistan in addition to offering training resources in Afghanistan. DOD's draft instruction for biometrics emphasizes the importance of training leaders in the effective employment of biometrics collection, but existing training does not instruct military leaders on (1) the effective use of biometrics, (2) selecting the appropriate personnel for biometrics collection training, and (3) tracking personnel who have been trained in biometrics collection to effectively staff biometrics operations. Absent this training, military personnel are limited in their ability to collect high-quality biometrics data to better confirm the identity of enemy combatants.

Several factors during the transmission process limit the use of biometrics in Afghanistan. Among them is unclear responsibility for the completeness and accuracy of biometrics data during their transmission. As a result, DOD cannot expeditiously correct data transmission issues as they arise, such as the approximately 4,000 biometrics collected from 2004 to 2008 that were separated from their associated identities. Such decoupling renders the data useless and increases the likelihood of enemy combatants going undetected within Afghanistan and across borders. Factors affecting the timely transmission of biometrics data include the biometrics architecture with multiple servers, mountainous terrain, and mission requirements in remote areas. These factors can prevent units from accessing transmission infrastructure for hours to weeks at a time. The DOD biometrics directive calls for periodic assessments, and DOD is tracking biometrics data transmission time in Afghanistan, but DOD has not determined the viability and cost-effectiveness of reducing transmission time.

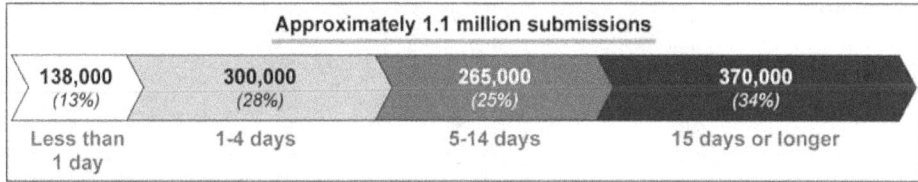

Timeliness of the Biometrics Transmission Process from October 2009 through October 2011

Source: GAO analysis of DOD data.

Lessons learned from U.S. military forces' experiences with biometrics in Afghanistan are collected and used by each of the military services and U.S. Special Operations Command. Military services emphasize the importance of using lessons learned to sustain, enhance, and increase preparedness to conduct future operations, but no requirements exist for DOD to disseminate existing biometrics lessons learned across the department.

United States Government Accountability Office

Contents

Letter		1
	Background	5
	DOD Conducts Biometrics Training, but Its Training for Leaders Does Not Fully Support Warfighter Use of Biometrics	12
	Biometrics Collections Occur across Afghanistan, but Several Factors during the Transmission Process Limit the Effectiveness of Biometrics Data	16
	Biometrics Lessons Learned Are Collected, but Not Disseminated across DOD	25
	Conclusions	26
	Recommendations for Executive Action	26
	Agency Comments	27
Appendix I	Scope and Methodology	30
Appendix II	GAO Contact and Staff Acknowledgments	34
Related GAO Products		35

Tables

	Table 1: Summary of DOD Biometrics Roles & Responsibilities	7
	Table 2: DOD Organizations Visited	30

Figures

	Figure 1: Example of Biometrics Used to Identify an Enemy Combatant	2
	Figure 2: Key DOD Biometrics Organizations	6
	Figure 3: Biometrics Collection Devices in Use	10
	Figure 4: Responsibility for Biometrics Data	17
	Figure 5: Timeliness of the Biometrics Transmission Process from October 2009 through October 2011	19
	Figure 6: Biometrics Architecture in Afghanistan	23

Abbreviations

ABIS   Automated Biometric Identification System
DOD   Department of Defense

This is a work of the U.S. government and is not subject to copyright protection in the United States. The published product may be reproduced and distributed in its entirety without further permission from GAO. However, because this work may contain copyrighted images or other material, permission from the copyright holder may be necessary if you wish to reproduce this material separately.

United States Government Accountability Office
Washington, DC 20548

April 23, 2012

Congressional Requesters

The long-term strategic success in counterinsurgency operations—such as Operation Enduring Freedom in Afghanistan—involves separating enemy combatants from innocent civilians in the general population. U.S. military forces in Afghanistan are using biometrics to identify enemy combatants and link individuals to events such as improvised explosive device detonations. Biometrics and related biometrics-enabled intelligence[1] are decisive, nonlethal battlefield capabilities that deny enemy combatants the necessary anonymity to hide and strike at will. From 2004 to 2011, U.S. military forces collected biometrics data in the form of over 1.6 million enrollments[2] involving more than 1.1 million persons in Afghanistan, and used biometrics to successfully identify approximately 3,000 known enemy combatants.

Biometrics is the measurement and analysis of a person's unique physical or behavioral characteristics, such as fingerprints or written signature recognition, which can be used to verify personal identity.[3] In Afghanistan, the U.S. military is using more than 7,000 electronic devices to collect biometrics data in the form of fingerprints, iris scans, and facial photographs. Fingerprints are made up of the minute ridge formations and patterns found on an individual's fingertips; iris scans are the pattern of a person's irises, which are muscles that control the amount of light that enters the eyes; and facial photographs are images that identify the location, shape, and spatial relationships of facial landmarks such as eyes, nose, lips, and chin. U.S. forces are using these three modalities to confirm a person's identity. When used in combination, the likelihood of matching biometrics data to a unique individual is substantially increased.

---

[1] Intelligence information associated with biometrics data that matches a specific person or unknown identity to a place, activity, device, or weapon and facilitates individual targeting, reveals movement patterns, and confirms claimed identity.

[2] To biometrically enroll an individual is to create and store a record that includes biometrics data and typically non-biometrics data, including intelligence information such as interrogation reports.

[3] Biometrics is a component of identity management—generally understood as the management of personal identity information. Examples of non-biometrics personal identity information include a person's name, Social Security number, and date of birth.

All biometrics data collected in Afghanistan are to be transmitted to the Department of Defense's (DOD's) Automated Biometric Identification System (ABIS) in West Virginia, where they are stored and used to identify enemy combatants by comparing and matching against previously collected biometrics data. Figure 1 is an example of how biometrics was used to identify an enemy combatant in Afghanistan.

Figure 1: Example of Biometrics Used to Identify an Enemy Combatant

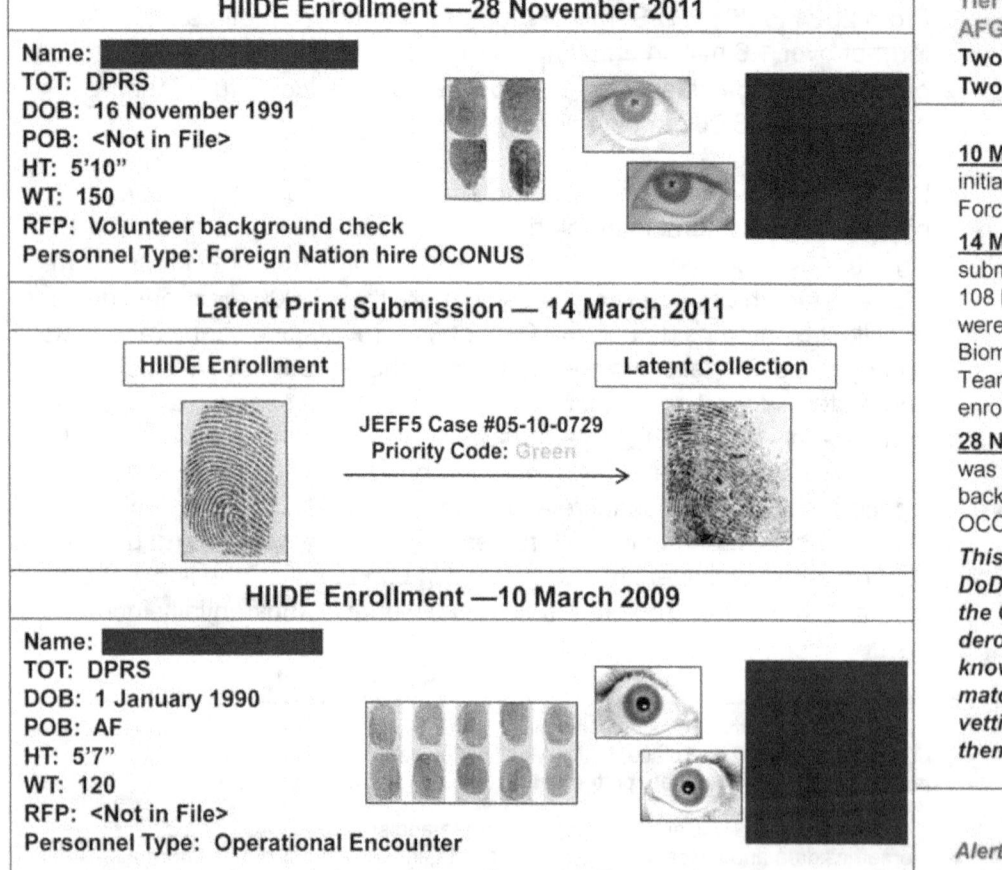

Source: Biometrics Identity Management Agency.

Note: Names and photographic information have been redacted for privacy purposes. The match is based on the biometrics information; the biographical information may vary depending on what the subject has provided.

In this example, in March 2009, the individual was initially biometrically enrolled during an operational encounter. In March 2011, this individual's latent fingerprints[4] were extracted from evidence and submitted to DOD's authoritative biometrics database—ABIS—where they were matched to the prior enrollment. In November 2011, the individual was voluntarily enrolled as a host nation hire and biometrically linked to the latent fingerprint submission from 8 months earlier.

The Secretary of the Army was designated the DOD Executive Agent for DOD Biometrics by Congress in July 2000 and assigned responsibility for leading and coordinating all DOD biometrics programs.[5] To assist him in carrying out his Executive Agent responsibilities, the Secretary of the Army designated the Biometrics Task Force in April 2008 to serve as the DOD Executive Manager for DOD Biometrics. In March 2010, the Biometrics Task Force was renamed the Biometrics Identity Management Agency. As the Executive Manager for DOD Biometrics, the Biometrics Identity Management Agency is charged with coordinating the department's efforts to program, integrate, and synchronize biometrics technologies and capabilities across the four military services and the nine unified combatant commands. The Biometrics Identity Management Agency also is responsible for developing and maintaining policies and procedures for the collection, processing, and transmission of biometrics data. To date, neither the Army nor the Marine Corps has institutionalized biometrics as a formal program of record, which would make biometrics a permanent capability. However, the Special Operations Command has a sensitive site exploitation program of record that includes biometrics.

To date, we have issued three reports in response to your two requests related to DOD biometrics. Our first two reports in 2008 focused on DOD's management of its biometrics activities and on the need for clearer guidance for the collection and sharing of biometrics data.[6] Our third report, issued in 2011, examined DOD's development of biometrics

---

[4] Latent fingerprints are images left on a surface touched by a person.

[5] Pub. L. No. 106-246, Division B, § 112 (2000).

[6] GAO, *Defense Management: DOD Needs to Establish Clear Goals and Objectives, Guidance, and a Designated Budget to Manage Its Biometric Activities*, GAO-08-1065 (Washington, D.C.: Sep. 26, 2008) and GAO, *Defense Management: DOD Can Establish More Guidance for Biometrics Collection and Explore Broader Data Sharing*, GAO-09-49 (Washington, D.C.: Oct. 15, 2008).

standards and the department's interagency biometrics information sharing efforts.[7] As agreed with your offices, given the continued importance of biometrics in military combat operations in Afghanistan, this report examines the extent to which (1) DOD's biometrics training supports warfighter use of biometrics, (2) DOD is effectively collecting and transmitting biometrics data, and (3) DOD has developed a process to collect and disseminate biometrics lessons learned.

Of the more than 1.6 million biometrics enrollments completed by U.S. military forces in Afghanistan, the Army collected approximately 86 percent, the Marine Corps collected approximately 11 percent, and U.S. Special Operations Command collected approximately 2 percent.[8] Therefore, the focus of this report is on the Army, and to a lesser extent the Marine Corps and Special Operations Command, biometrics collection efforts in Afghanistan.[9] To determine the extent to which DOD's biometrics training supports warfighter use of biometrics, we reviewed relevant Army, Marine Corps, U.S. Central Command, and Special Operations Command training policies and guidance; interviewed DOD officials in the United States and Afghanistan; and observed several training courses. DOD also has implemented biometrics training for intelligence analysts, but it was not included in our review. To determine the extent to which DOD is effectively collecting and transmitting biometrics data, we reviewed relevant policies and guidance and interviewed officials from Office of the Secretary of Defense, Biometrics Identity Management Agency, Army, Marine Corps, Central Command, and Special Operations Command. We observed the collection and transmission of biometrics data at installation entry points and other locations in Afghanistan. We also toured DOD laboratories in Afghanistan that exploit evidence obtained from the battlefield for biometrics data and interviewed officials at these sites. To determine the extent to which DOD has developed a process to collect and disseminate biometrics lessons learned, we reviewed relevant DOD guidance and interviewed officials

---

[7] GAO, *Defense Biometrics: DOD Can Better Conform to Standards and Share Biometric Information with Federal Agencies*, GAO-11-276 (Washington, D.C.: Mar. 31, 2011).

[8] While forces under Special Operations Command collect a small proportion of biometrics in Afghanistan, the collections they do make are often from specific, high-value individuals. Due to rounding, percentages do not total 100 percent.

[9] We met with Navy and Air Force officials to confirm that they have collected comparatively few biometrics enrollments in Afghanistan.

from the Army and Marine Corps lessons learned centers as well as knowledgeable Joint Staff, military service, Central Command, Special Operations Command, and Biometrics Identity Management Agency officials. More detailed information on our scope and methodology can be found in appendix I of this report.

We conducted this performance audit from May 2011 through April 2012, in accordance with generally accepted government auditing standards. Those standards require that we plan and perform the audit to obtain sufficient, appropriate evidence to provide a reasonable basis for our findings and conclusions based on our audit objectives. We believe that the evidence obtained provides a reasonable basis for our findings and conclusions based on our audit objectives.

Background

DOD Biometrics Organization

Defense biometrics activities involve a number of military services, commands, and offices across the department. Figure 2 depicts the relationship among several of the key DOD biometrics organizations.

Figure 2: Key DOD Biometrics Organizations

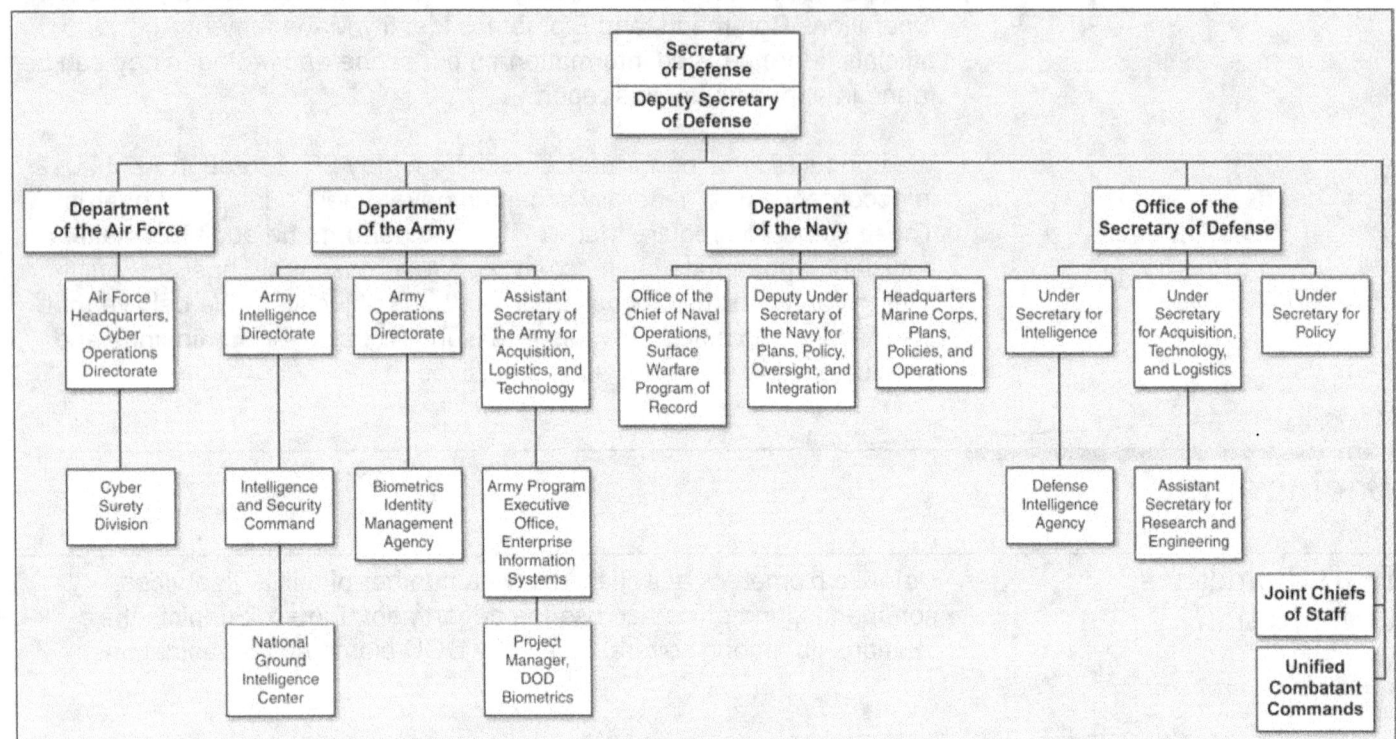

Source: GAO analysis of DOD information.

Roles and responsibilities for defense biometrics activities are explained in DOD's 2008 biometrics directive, and summarized in table 1.[10]

---

[10] DOD Directive 8521.01E, *Department of Defense Biometrics* (Feb. 21, 2008).

Table 1: Summary of DOD Biometrics Roles & Responsibilities

DOD entity	Roles and responsibilities for DOD biometrics
Office of the Under Secretary of Defense for Acquisition, Technology, and Logistics, Office of the Assistant Secretary of Defense for Research & Engineering[a]	As the Principal Staff Assistant • oversee biometrics programs, initiatives, technologies, and policies • oversee interagency coordination • oversee biometrics funding
Office of the Under Secretary of Defense for Policy	• Support DOD biometrics with policy development, implementation, and oversight • Coordinate with the Biometrics Principal Staff Assistant on certain biometrics issues, such as certain acquisition programs and submissions
Office of the Secretary of the Army	As the Executive Agent • appoint an executive manager • appoint a program management office • coordinate biometrics requirements
Biometrics Identity Management Agency	As the Executive Manager[b] • provide biometrics research and technology support • develop policies, processes, and procedures for biometrics data and associated intelligence • manage authoritative biometrics database • act as the focal point for biometrics data sharing
Office of the Project Manager for DOD Biometrics	• develop, acquire, and field common biometrics systems to support military service and joint service requirements
Military Services and Special Operations Command	• coordinate biometrics strategies and requirements prior to acquisition • conform to approved DOD biometrics hardware and software architecture • comply with DOD-approved policies, standards, and processes for biometrics • coordinate with the Executive Agent to avoid program duplication • develop service-level biometrics training, direction, and implementation guidance • develop joint warfighting requirements[c] • develop resource requirements process for acquisition program objectives[c]
Combatant Commands	• develop joint warfighting requirements • develop resource requirements process for acquisition program objectives
Defense Intelligence Agency	• develop, maintain, and share a DOD biometrically enabled watchlist[d] • develop intelligence collection and analysis capabilities to incorporate information derived from biometrics

Source: GAO analysis of DOD information.

Notes:

[a]DOD Directive 8521.01E assigns responsibilities to the "Director, Defense Research and Engineering." The Director, Defense Research and Engineering has been redesignated as the Assistant Secretary of Defense for Research and Engineering. For purposes of this report, we use the new position title when referring to responsibilities or activities of both the current and prior office.

[b]By September 30, 2012, a number of Executive Manager responsibilities are expected to transfer to the Army's Office of the Provost Marshal General.

[c]Specific to Special Operations Command.

ᵈSince 2006, the Army's National Ground Intelligence Center has established and maintained the biometrically enabled watchlist. The 2008 DOD biometrics directive assigns this responsibility to the Director, Defense Intelligence Agency.

DOD is revising its biometrics directive based on, among other things, new requirements in the Ike Skelton National Defense Authorization Act for Fiscal Year 2011.[11] Office of the Secretary of Defense officials said that they plan to issue the revised biometrics directive in the fall of 2012. The office had started to draft an implementing instruction for biometrics based on the 2008 directive but suspended this effort pending issuance of the updated directive. According to DOD officials, the implementing instruction is expected to contain a more detailed description of roles and responsibilities based upon the revised directive.

To oversee biometrics activities in Afghanistan, Central Command established Task Force Biometrics in 2009. According to the *Commander's Guide to Biometrics in Afghanistan*, Task Force Biometrics assists commands with integrating biometrics into their mission planning, trains individuals on biometrics collection, develops biometrics-enabled intelligence products, and manages the biometrically enabled watchlist for Afghanistan that contains the names of more than 33,000 individuals.[12] This watchlist is a subset of the larger biometrically enabled watchlist managed by the National Ground Intelligence Center. Additionally, according to Army officials, the Army established the Training and Doctrine Command Capabilities Manager for Biometrics and Forensics with responsibilities for ensuring that user requirements are considered and incorporated in Army policy and doctrine involving biometrics. Further, the Army gave its Intelligence Center of Excellence responsibilities for developing and implementing biometrics training, doctrine, education, and personnel.

Biometrics collection process in Afghanistan

U.S. forces are collecting biometrics data on non-U.S. persons[13] in Afghanistan at roadside checkpoints, base entry control points, and

---

[11] Pub. L. No. 111-383, § 121 (2011).

[12] U.S. Army, *Commander's Guide to Biometrics in Afghanistan: Observations, Insights, and Lessons*, Center for Army Lessons Learned (April 2011).

[13] Non-U.S. persons are individuals who are neither U.S. citizens nor aliens lawfully admitted into the United States for permanent residence.

during patrols and other missions. U.S. forces use three principal biometrics collection devices to enroll individuals.

- *The Biometrics Automated Toolset*: Consists of a laptop computer and separate peripherals for collecting fingerprints, scanning irises, and taking photographs. The Toolset system connects into any of the approximately 150 computer servers geographically distributed across Afghanistan that store biometrics data. The Toolset system is used to identify and track persons of interest and to document and store information, such as interrogation reports, about those persons. This device is primarily used by the Army and Marine Corps to enroll and identify persons of interest.
- *The Handheld Interagency Identity Detection Equipment*: Is a self-contained handheld biometrics collection device with an integrated fingerprint collection surface, iris scanner, and camera. The Handheld Interagency Identity Detection Equipment connects to the Biometrics Automated Toolset system to upload and download biometrics data and watchlists. This device is primarily used by the Army and Marine Corps.
- *The Secure Electronic Enrollment Kit*: Is a self-contained handheld biometrics collection device with a built-in fingerprint collection surface, iris scanner, and camera. Additionally, the Secure Electronic Enrollment Kit has a built-in keyboard to facilitate entering biographical and other information about individuals being enrolled. The Kit is used primarily by the Special Operations Command, although the Army and Marine Corps have selected the Kit as the replacement biometrics collection device for the Handheld Interagency Identity Detection Equipment.

The Biometrics Automated Toolset, Handheld Interagency Identity Detection Equipment, and Secure Electronic Enrollment Kit collection devices are shown in figure 3.

Figure 3: Biometrics Collection Devices in Use

Biometrics Automated Toolset (BAT)

Handheld Interagency Identity Detection Equipment (HIIDE)

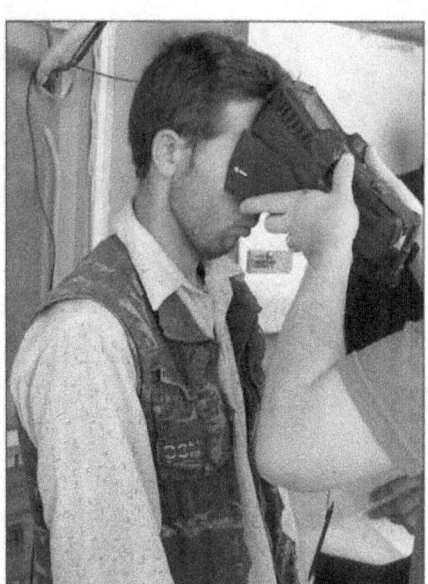
Secure Electronic Enrollment Kit (SEEK)

Sources: GAO (BAT and SEEK photos); DOD (HIIDE photo).

U.S. forces in Afghanistan collect biometrics data and search for a match against the Afghanistan biometrically enabled watchlist that is stored on the biometrics collection devices in order to identify persons of interest. Soldiers and Marines connect their biometrics collection devices to the Afghanistan Biometrics Automated Toolset system's architecture, at which point the data are transmitted and replicated[14] through a series of computer servers in Afghanistan to the ABIS database in West Virginia. Special operations forces have a classified and an unclassified Web-based portal that they use to transmit biometrics data directly from their collection devices to the ABIS database in West Virginia. Biometrics data obtained during the enrollment using the biometrics collection devices are searched against previously collected biometrics records in the

---

[14] Data replication is the process of sharing data so as to ensure consistency between redundant resources, such as software or hardware components, to improve reliability or accessibility.

Afghanistan biometrically enabled watchlist, and in some cases the Biometrics Automated Toolset servers, before searching against stored biometrics records and latent fingerprints stored in ABIS. Match/no match watchlist results are reported to Task Force Biometrics and other relevant parties. The biometrics data collected during the enrollment are retained in ABIS for future matching by DOD.

Once collected, biometrics data and associated information are evaluated by intelligence analysts to link a person with other people, events, and information. This biometrics-enabled intelligence is then used to identify persons of interest, which can result in his or her inclusion on the biometrically enabled watchlist. The biometrically enabled watchlist for Afghanistan contains five levels, and according to the level of assignment, an individual who is encountered after his or her initial enrollment will be detained, questioned, denied access to U.S. military bases, disqualified from training or employment, or tracked to determine his or her activities and associations.

Other Biometrics Databases Operated by U.S. Federal Agencies

In addition to DOD, the Federal Bureau of Investigation and the Department of Homeland Security collect and store biometrics data to identify persons of interest. The Federal Bureau of Investigation uses its biometrics system for law enforcement purposes. The Department of Homeland Security uses its biometrics system for border security, naturalization, and counterterrorism purposes, as well as for visa approval in conjunction with the Department of State. While the three biometrics organizations are able to share information, the biometrics databases operate independently from one another, as we have noted in our March 2011 report.[15]

---

[15] See GAO-11-276.

DOD Conducts Biometrics Training, but Its Training for Leaders Does Not Fully Support Warfighter Use of Biometrics

DOD has trained thousands of personnel on the collection and transmission of biometrics data since 2004; however, training for leaders does not fully support warfighter use of biometrics because it does not instruct unit commanders and other military leaders on (1) the effective use of biometrics, (2) selecting the appropriate personnel for biometrics collection training, and (3) tracking personnel who have been trained in biometrics collection to effectively staff biometrics operations.

DOD Conducts Biometrics Training

The Army, Marine Corps, and Special Operations Command have trained thousands of personnel on the use of biometrics prior to their deployment to Afghanistan over the last 8 years. This training includes the following:

- *Army*: Offers classroom training at its three combat training centers at Fort Polk, Louisiana; Fort Irwin, California; and Hohenfels, Germany as well as home station training teams and mobile training teams that are available to travel and train throughout the United States as needed. In addition, the Army is developing virtual-based training software to supplement its classroom training efforts.
- *Marine Corps*: Offers classroom training at its training centers at Camp Pendleton, California and Camp LeJeune, North Carolina as well as simulation training at Twentynine Palms, California.
- *Special Operations Command*: Offers classroom and simulation training at Fort Bragg, North Carolina.

Moreover, the military services and Special Operations Command have mobile training teams in Afghanistan to provide biometrics training to personnel during their deployment. Additionally, the military services rely on personnel who have been trained in biometrics prior to deployment to train others while deployed.

The Office of the Secretary of Defense, the military services, and Central Command each has emphasized in key documents the importance of training. The 2008 DOD directive, which was issued by the Under Secretary of Defense for Acquisition, Technology, and Logistics 4 years after biometrics collection began in Afghanistan, emphasizes the importance of biometrics training, including the need for component-level guidance to ensure training is developed as required.[16] The Office of the

---

[16] DOD Directive 8521.01E, *Department of Defense Biometrics* (Feb. 21, 2008).

Assistant Secretary of Defense for Research and Engineering[17] subsequently drafted an implementing instruction that includes guidance for the establishment of training programs designed to enable DOD units and leaders to effectively employ biometrics collection capabilities and utilize biometrically enabled watchlists.[18] As noted earlier, this instruction will not be issued until the Office of the Under Secretary of Defense for Acquisition, Technology, and Logistics reissues the biometrics directive, potentially in the fall of 2012. Both the Army and Central Command have issued guidance that requires soldiers to be trained prior to deployment.[19] Additionally, an Army regulation on training says that first and foremost, training must establish tasks, conditions, and standards to prepare units to perform their missions.[20] Similarly, a Marine Corps order on training states that units focus their training effort on those missions and tasks to which they can reasonably expect to be assigned in combat.[21] This Office of the Secretary of Defense, Army, Marine Corps, and Central Command guidance underscores the importance of biometrics training.

Biometrics Training for Leaders Does Not Provide Detailed Instruction on the Effective Use and Management of Biometrics

DOD's draft instruction for biometrics emphasizes the importance of training leaders in the effective employment of biometrics. However, existing biometrics training for leaders does not instruct unit commanders and other military leaders on (1) the effective use of biometrics, (2) selecting the appropriate personnel for biometrics collection training, and (3) tracking personnel who have been trained in biometrics collection to effectively staff biometrics operations. When leaders are not invested in the importance of biometrics as a tool for identifying enemy combatants, the warfighters serving with them may be unaware of the value of biometrics because their leaders have not conveyed to them the importance of biometrics. Moreover, existing biometrics training for

---

[17] The Office of the Assistant Secretary of Defense for Research and Engineering is within the Office of the Under Secretary of Defense for Acquisition, Technology, and Logistics.

[18] DOD Instruction 8521.bb, *DOD Instruction on Biometrics* (draft as of October 2011).

[19] U.S. Army Forces Command, *Predeployment Training Guidance In Support Of Combatant Commands,* (September, 2011), and U.S. Central Command, Fragmentary Order 09-1700: *Central Command Theater Training Requirements* (March 2011).

[20] Army Regulation 350.1. *Army Training and Leader Development* (Sep. 4, 2011).

[21] Marine Corps Order P3500.72A, *Marine Corps Ground Training and Readiness Program* (Apr. 18, 2005).

leaders limits the ability of military personnel to collect higher quality biometrics enrollments[22] to better confirm the identity of enemy combatants through biometrics.

The military services and Special Operations Command have developed biometrics training for leaders to varying degrees, but the existing training does not communicate how leaders can effectively use biometrics in their mission planning. As noted in *Army Training Needs Analysis for Tactical Biometrics Collection Devices* issued in March 2010, a majority of the Army's unit commanders were unaware of how biometrics collection contributes to identifying enemy combatants, and that failure to address biometrics in training for leaders hampers force protection measures. This analysis also stated that Army leaders need to train on how and when to incorporate biometrics in mission planning and how to subsequently deploy soldiers to use biometrics systems. As a result of the analysis, the Army developed a 1-hour briefing for unit commanders and other senior officials, but according to an Army training official, it is voluntary training provided by mobile training teams and not a part of the Army's formal, required training for leaders. Furthermore, even if leaders take the briefing, they may still not be fully aware of the importance and use of biometrics in combat missions because the briefing focuses primarily on operating biometrics devices for collecting and transmitting biometrics data and not on the value of biometrics' contribution to identifying enemy combatants. In addition, neither the Marine Corps nor the Special Operations Command has incorporated training for leaders into its biometrics training efforts. A Marine Corps official told us that biometrics training for leaders will not be developed until the Marine Corps finalizes its *Marine Corps Identity Operations Strategy 2020*, which will establish biometrics as a fully integrated capability. Officials at Special Operations Command said that while they offer biometrics training for the warfighter, they do not have dedicated biometrics training for leaders.

Existing Army biometrics training for leaders also does not stress the importance of (1) selecting appropriate personnel for biometrics training, and (2) tracking which personnel have completed biometrics training prior to deployment. These two omissions likely contribute to less effective biometrics operations. For example, the Army found that unit

---

[22] Higher quality biometrics enrollments—that is, enrollments that contain fingerprints and iris scans that have a large number of clear and complete data points—result in improved matching performance by DOD ABIS and reduced search times.

commanders frequently made inappropriate choices regarding which soldiers should attend biometrics collection training prior to deploying to Afghanistan (e.g., vehicle drivers) compared to other occupations (e.g., military police) who are more likely to utilize biometrics in operations. Similarly, a Marine Corps official told us that commanders have selected personnel for biometrics missions who were never identified for predeployment training, including, in one instance, musicians. With respect to tracking personnel with biometrics training, the Army requires unit commanders to document soldiers' completion of unit-level training in the Army's Digital Training Management System.[23] However, the Army Audit Agency reported in March 2011 that units routinely do not use the system to document training—biometrics or otherwise—and Army biometrics officials with whom we spoke during the course of our review were unaware of this system or any other mechanism to track completion of biometrics training.[24] Similarly, the Marine Corps does not have a tracking mechanism to identify personnel trained in biometrics prior to deployment. A Marine Corps training official said that because they have not developed biometrics doctrine and training guidance, biometrics training is not tracked. Consequently, Army and Marine Corps unit commanders in Afghanistan do not have accurate information on which and how many of their personnel have received training for conducting biometrics operations. This lack of accurate information impedes unit commanders' ability to assess whether they have sufficient expertise among their personnel to effectively staff biometrics operations.

---

[23] Army Regulation 350-1, *Army Training and Leader Development* (Sep. 4, 2011).

[24] U.S. Army Audit Agency, *Digital Training Management System,* Report No. A-2011-0075-FFT (Mar. 10, 2011).

Biometrics Collections Occur across Afghanistan, but Several Factors during the Transmission Process Limit the Effectiveness of Biometrics Data

Since 2004, U.S. forces in Afghanistan have collected biometrics from more than 1.2 million individuals with approximately 3,000 successful matches to enemy combatants, but factors during the transmission process limit biometrics' timely identification of enemy combatants using biometrics.[25]

Responsibility for Biometrics Data during the Transmission Process Is Unclear

Every week, thousands of biometrics enrollments are collected in Afghanistan and transmitted to ABIS in West Virginia; however, responsibility for assuring the completeness and accuracy of the biometrics data during the transmission process is unclear. According to the DOD biometrics directive, the Executive Manager for DOD Biometrics is responsible for developing and maintaining policies and procedures for the collection, processing, and transmission of biometrics data.[26] However, no policy has been articulated that assigns responsibility for maintaining the completeness and accuracy of biometrics data during the transmission process. In addition, the *Standards for Internal Control in the Federal Government* state that (1) controls should be installed at an application's interfaces with other systems to ensure that all inputs are received and are valid, and that outputs are correct and properly distributed; and (2) key duties and responsibilities are divided among different people to reduce the risk of error.[27]

As shown in figure 4, the warfighter has responsibility for the biometrics data from collection to the point of submission into the Biometrics

---

[25] By transmission process we mean the time period between when warfighters submit biometrics enrollments into the Biometrics Automated Toolset system and when the data are received by ABIS for matching and storage.

[26] DOD Directive 8521.01E, §E4.4 (2008).

[27] GAO, *Standards for Internal Control in the Federal Government*, GAO/AIMD-00-21.3.1 (Washington, D.C.: November 1999).

Automated Toolset system, and the Biometrics Identity Management Agency assumes responsibility for the biometrics data once they are received by ABIS. The Project Manager for DOD Biometrics has responsibility for the physical infrastructure of the Biometrics Automated Toolset system. DOD officials we spoke with were unable to identify who has responsibility for the completeness and accuracy of the biometrics data during the transmission process. Specifically, officials from Central Command stated that it owns the biometrics data, but the Project Manager for DOD Biometrics is responsible for their completeness and accuracy. Officials from the Project Manager for DOD Biometrics told us that it is not responsible for the completeness and accuracy of the biometrics data.

Figure 4: Responsibility for Biometrics Data

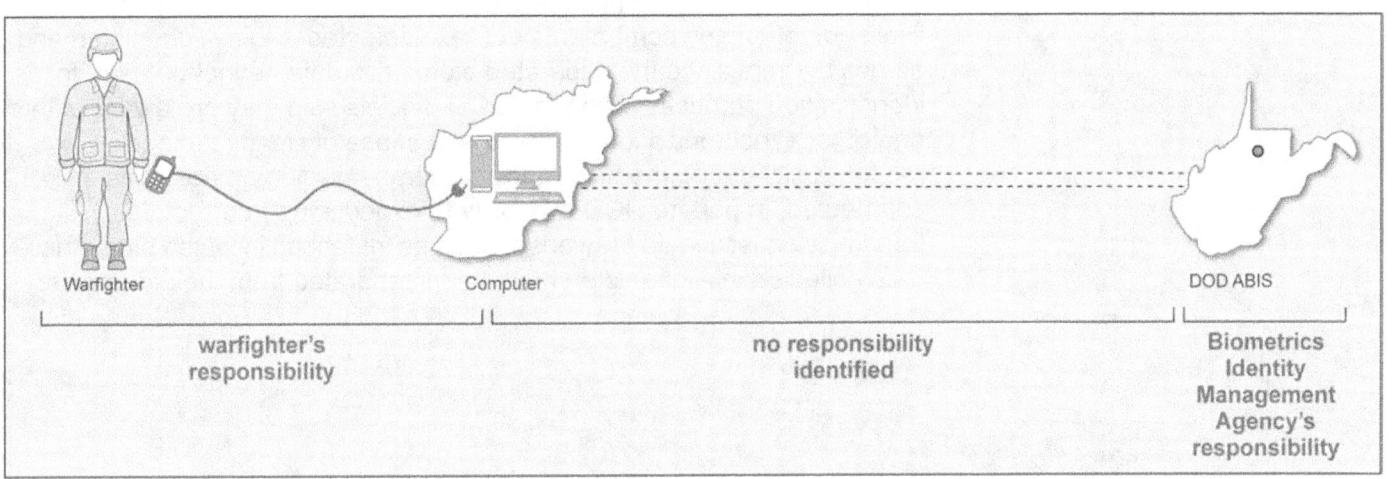

Source: GAO analysis of DOD information.

In some cases, issues during the transmission process have surfaced impacting the completeness and accuracy of the biometrics data. Specifically, data synchronization[28] issues led Central Command to issue

---

[28] Data synchronization is the process of establishing consistency among data from a source to a target data storage and vice versa and the continuous harmonization of such data over time. Hence, biometrics systems which are not properly synchronized could result in de-linked data, rendering them unreliable.

an urgent requirement in September 2009[29] to improve data synchronization to avoid further hindering DOD's ability to transfer biometrics data; however, the urgent needs statement was rescinded[30] following more than a year of inaction without improvements having been made in order to reallocate funding towards the Last Tactical Mile pilot project.[31] This issue has continued to impact the completeness and accuracy of biometrics data. For example, during the Last Tactical Mile pilot project in summer 2011, Army officials found that of the more than 33,000 people on the Afghanistan biometrically enabled watchlist, approximately 4,000 biometrics collected from 2004 to 2008 had become separated from their associated identities and 1,800 remained separated as of October 2011. Officials stated that the separated data were most likely due in part to synchronization issues during the data transmission process. This decoupling of an individual from his or her associated biometrics data undermines the utility of biometrics by increasing the likelihood of enemy combatants going undetected within Afghanistan and across borders since the separated biometrics data cannot be used for identification purposes. Although DOD officials said they are aware of this and other synchronization issues, the absence of clearly defined responsibility during the biometrics data transmission process has contributed, in part, to DOD's inability to expeditiously correct data transmission issues as they arise, such as instances in which biometrics data collected in Afghanistan have been separated from their identities.

---

[29] U.S. Forces-Afghanistan, *Biometrics Data Synchronization Joint Urgent Operational Need Statement* CC-0385 (Sep. 22, 2009).

[30] Memorandum for Central Command Commander, *U.S. Forces-Afghanistan Request to Rescind Joint Urgent Operational Need Statement CC-0385 for Biometrics Synchronization* (May 18, 2011).

[31] The Last Tactical Mile pilot project originated as part of a joint urgent operational need to utilize wireless infrastructure to transmit biometrics data from handheld biometrics collection devices to a wireless communications tower, with a capability of providing match/no match responses to the warfighter with watchlist and latent fingerprint matches within 2 to 5 minutes of transmission.

Other Factors Occurring In the Biometrics Transmission Process Affect Timely Enemy Combatant Detection with Biometrics

Several factors in transmitting biometrics data from Army and Marine Corps forces affect DOD's ability to identify and capture enemy combatants with biometrics in a timely manner. The transmission process for biometrics data involves a unit's submission of collected enrollments, matching in ABIS, and a match/no match response back to the unit. From the time data are submitted, the transmission process can take from less than 1 day to 15 days or more to complete, as shown in figure 5. However, the design specifications for the Biometrics Automated Toolset system for Afghanistan state that biometrics data should transmit from the point of data submission to ABIS within 4 hours.

Figure 5: Timeliness of the Biometrics Transmission Process from October 2009 through October 2011

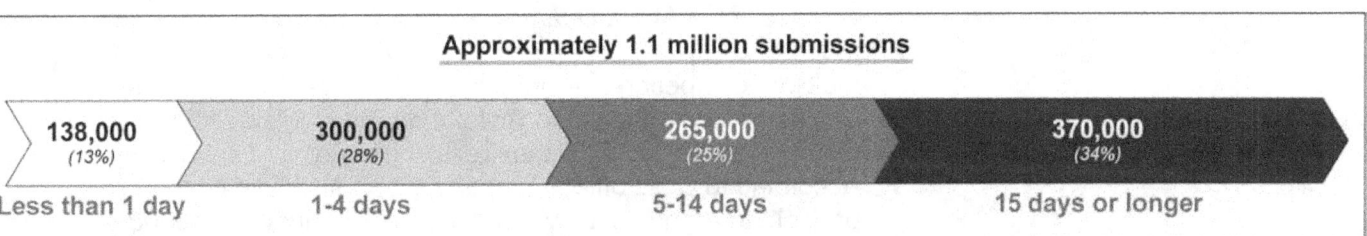

Source: GAO analysis of DOD data.

In contrast, according to officials from the Biometrics Identity Management Agency, once enrollments are received by ABIS, the time it takes to match the data and transmit a response to the National Ground Intelligence Center for intelligence analysis, and ultimately back to the unit that performed the biometrics enrollment in Afghanistan, averages 22 minutes.[32]

Multiple factors contribute to the time it takes to transmit biometrics data from the warfighter to ABIS, and back. These factors include:

- *Biometrics architecture*: The Biometrics Automated Toolset system's architecture constructed for use in Afghanistan requires biometrics submissions to be replicated sequentially across multiple computer servers before reaching ABIS. As noted in figure 5, biometrics data on the Biometrics Automated Toolset system's architecture can take

---

[32] Special operations forces receive match/no match responses directly from the Biometrics Identity Management Agency.

more than 2 weeks to transmit from Afghanistan to ABIS. However, DOD is unclear on how the number of servers correlates to transmission timeliness.

- *Geographic challenges to connectivity*: The mountainous terrain in Afghanistan's northern regions highlights the limited ability of U.S. forces to transmit biometrics data within the country. For example, under optimal conditions (i.e., flat terrain), wireless transmission, such as that used in the Last Tactical Mile pilot project, is capable of transmitting biometrics data up to approximately 50 miles.[33] However, wireless transmission requires line-of-sight from a handheld biometrics device to a wireless tower, which would necessitate acquiring and erecting many towers to cover a relatively small geographic area. DOD is still evaluating the viability of expanding the pilot project in Afghanistan.
- *Multiple, competing demands for communications infrastructure*: Multiple, competing demands for communications infrastructure by U.S. forces in Afghanistan limit bandwidth available to transmit biometrics data to ABIS, thus resulting in delayed submissions. According to DOD officials, available bandwidth is a continuing problem in Afghanistan, which limits the amount and speed of information transmitted within or outside of Afghanistan. DOD has increased bandwidth capacity in Afghanistan over the years, but new military capabilities add to the demand for additional bandwidth.
- *Mission requirements*: According to the *Commander's Guide to Biometrics in Afghanistan*, forces should submit enrollments to ABIS within 8 hours of completion of a mission; however, missions can keep units operating in remote areas away from biometrics transmission infrastructure for weeks at a time.[34] While on missions, a unit's biometrics collection devices have a preloaded Afghanistan biometrically enabled watchlist and are typically updated weekly, but again, mission requirements can delay updating these devices with the most current watchlist.

DOD has pursued two key efforts to reduce the time it takes to transmit biometrics data in Afghanistan outside of the Biometrics Automated Toolset system: communication satellites used by special operations

---

[33] The Last Tactical Mile pilot project utilizes mobile transmission points and fixed wireless towers capable of transmitting biometrics data approximately 50 miles. In situations with reduced line-of-sight, transmission distances decrease.

[34] U.S. Army, *Commander's Guide to Biometrics in Afghanistan: Observations, Insights, and Lessons*, Center for Army Lessons Learned (April 2011).

forces and the Last Tactical Mile pilot project. Special operations forces upload their biometrics enrollments to a dedicated classified or unclassified Web-based portal using communications satellites.[35] The Biometrics Identity Management Agency monitors the portal and accesses the enrollments therein to match against biometrics data stored in ABIS. In addition to fingerprint, iris, and facial biometrics, the Web-based portal supports the cataloguing and analysis of other biometric and nonbiometric evidence such as DNA, documents, and cell phone data. Match/no match responses are provided to the warfighter via the portal within 2 to 7 minutes, assuming satellite or other Internet connectivity is available.[36] Additionally, special operations forces can match individuals against a preloaded biometrically enabled watchlist on the handheld biometrics collection devices.

Central Command was responsible for conducting the Last Tactical Mile pilot project during 2011 to provide the warfighter with a rapid response time on biometrics data submissions. In the pilot project, matching is initially against a biometrically enabled watchlist stored on the warfighter's handheld device before searching against data stored in a stand-alone computer server in Afghanistan prior to transmission to ABIS—the authoritative database. The Last Tactical Mile pilot project originated as a joint urgent operational need[37] to utilize wireless infrastructure to transmit biometrics enrollments from a handheld biometrics collection device to a wireless communications tower.[38] A goal of the pilot project was to receive a match/no match response in 2 to 5 minutes against the biometrics data stored on the computer server in Afghanistan, including possible latent fingerprint matches. Army officials told us that expanding the Last Tactical Mile pilot project across all of Afghanistan would cost approximately $300 million, in large part due to the number of wireless

---

[35] Since the portal is Web-based and the biometrics collection devices are handheld, this approach can be used worldwide.

[36] The Biometrics Identity Management Agency prioritizes biometrics data submissions based upon user needs, with special operations forces receiving the highest priority. Therefore, some biometrics submissions may have longer response times.

[37] U.S. Forces-Afghanistan Joint Urgent Operational Need for Biometrics Last Tactical Mile Solution, CC-0434 (Jan. 4, 2011).

[38] In addition to direct transmission to the wireless communications tower, the handheld biometrics collection devices can wirelessly transmit to a mobile transmission point attached to a military vehicle that then relays the biometrics data to the wireless communications tower.

communications towers that would be necessary to provide connectivity across the mountainous terrain in the northern part of the country. DOD had not completed its evaluation of the Last Tactical Mile pilot project at the time of our review, and had not documented plans to utilize wireless infrastructure for biometrics in Afghanistan beyond the continued operation of the pilot project.

Figure 6 highlights the differences between the Biometrics Automated Toolset system's architecture used by the Army and the Marine Corps, the Web-based architecture used by Special Operations Command, and the architecture in the Last Tactical Mile pilot project.

Figure 6: Biometrics Architecture in Afghanistan

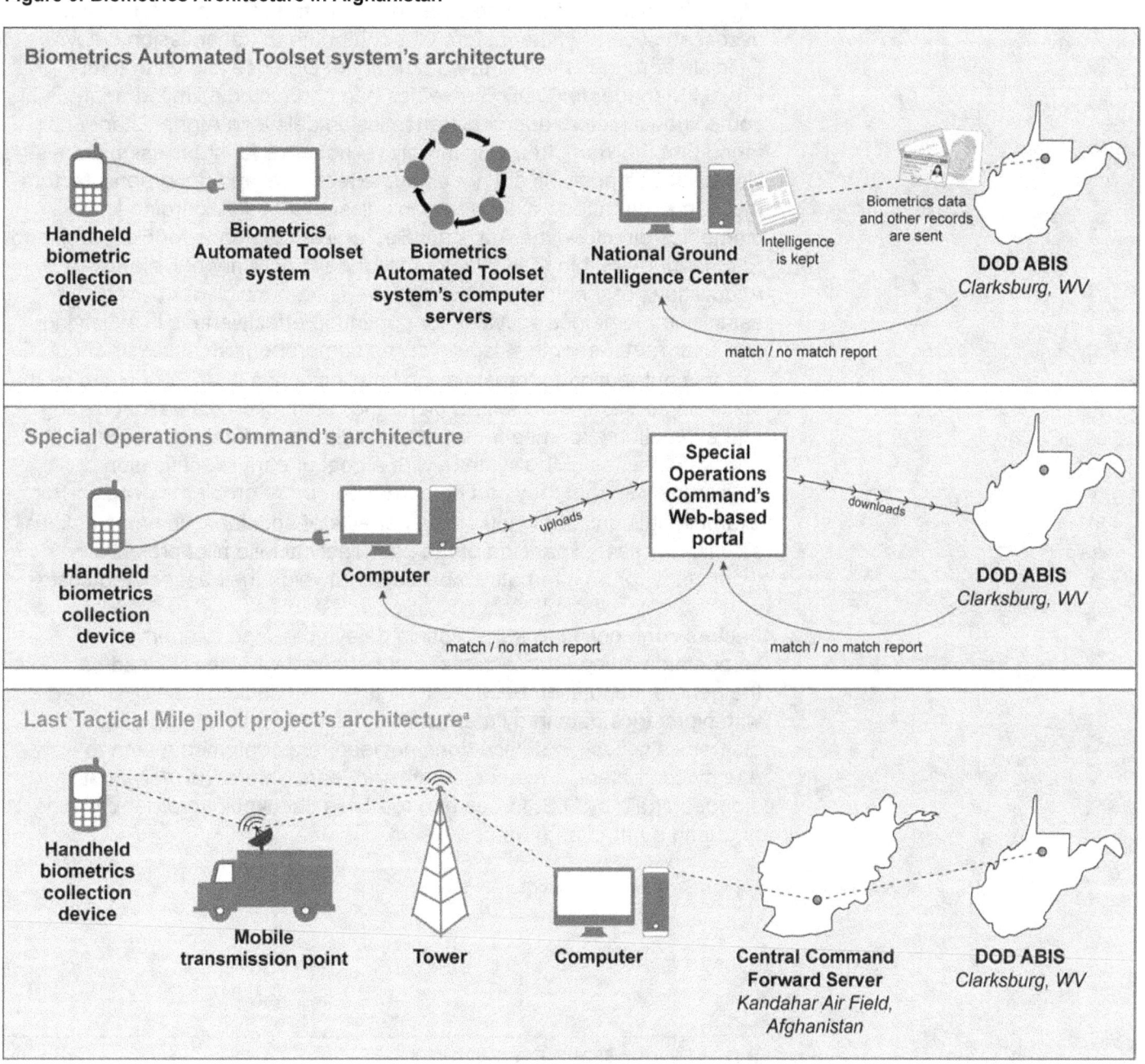

Source: GAO analysis of DOD information.

Note: [a]In the Last Tactical Mile pilot project, match/no match reports are initially provided to the warfighter from a biometrically enabled watchlist stored on the handheld collection device followed by the Central Command Forward Server for latent fingerprints and watchlist match results. Once biometrics data are received by ABIS, another match/no match report is generated back to the Central Command Forward Server based upon the full ABIS database.

Although DOD is tracking biometrics data transmission time in Afghanistan to facilitate timely responses to the warfighter, it has not assessed several of the factors that contribute to transmission delays. Officials from the Office of the Secretary of Defense told us that the Project Manager for DOD Biometrics had conducted a limited analysis of some known factors affecting transmission delays in Afghanistan, and found that the warfighter was largely responsible for submission delays. However, this analysis did not evaluate technical and geographic factors that can contribute to extended transmission times. According to the biometrics directive, the Assistant Secretary of Defense for Research and Engineering—within the Office of the Under Secretary of Defense for Acquisition, Technology, and Logistics—is responsible for periodically assessing biometrics activities for continued effectiveness in satisfying end-user requirements.[39] However, no comprehensive assessment of factors contributing to transmission timeliness has been conducted by this office. In addition, DOD's draft biometrics instruction states that testing and evaluation expertise must be employed to understand the strengths and weaknesses of the system, with a goal of early identification of deficiencies so that they can be corrected before problems occur.[40] For example, it is unclear whether the benefits of additional communications satellite access, expansion of the Last Tactical Mile pilot project's technology, or an alternative approach outweigh their associated costs.

Factors contributing to transmission delays can lead to enemy combatants going undetected and subsequently being released back into the general population because their identities could not be confirmed with biometrics data in a timely manner. If a watchlist stored on a biometrics collection device does not lead to a confirmed match to an enemy combatant, it may be months or years before the individual is stopped again by U.S. forces at a roadside checkpoint, border crossing, or during a patrol or another mission, if ever.

---

[39] DOD Directive 8521.01E, § 5.1.2.1 (2008).

[40] As noted, this instruction is in draft, and is not mandatory until issued. It is not planned for issuance until the Office of the Secretary of Defense reissues the biometrics directive, potentially in the fall of 2012.

Biometrics Lessons Learned Are Collected, but Not Disseminated across DOD

Lessons learned from U.S. forces' experiences with biometrics in Afghanistan are collected by and used within each of the military services and Special Operations Command, but those lessons are not disseminated across DOD. Army and Marine Corps guidance both emphasize using lessons learned to sustain, enhance, and increase preparedness to conduct current and future operations.[41] The Army, Marine Corps, and Special Operations Command each rely on their respective existing processes to collect lessons learned pertaining to biometrics to facilitate knowledge sharing. To collect lessons learned, the military services and Special Operations Command draw from a variety of sources, including through surveys administered to students and instructors during training, and through interviews with personnel who have recently returned from a deployment. These lessons learned are analyzed to identify opportunities to improve existing practices within the military services and Special Operations Command. For example, Army officials said that about 10 to 15 percent of the lessons learned it collects are subsequently identified as either best practices or issues that require further action to resolve.

DOD also uses informal processes to capture biometrics training-related lessons learned. For example, monthly teleconferences are held by and open to training representatives from Central Command's Task Force Biometrics and the Army to discuss biometrics training-related issues and experiences. However, this information is not disseminated across the department. Army biometrics officials told us that it would be advantageous to share biometrics lessons learned across the military services and combatant commands.

Currently, DOD has no requirement to disseminate biometrics lessons learned across the department. However, the unpublished DOD implementing instruction for biometrics that was drafted by the Office of the Assistant Secretary of Defense for Research and Engineering[42]

---

[41] Army Regulation 11-33, *Army Lessons Learned Program* (Oct. 17, 2006); Marine Corps Order 3504.1, *Marine Corps Lessons Learned Program (MCLLP) and the Marine Corps Center For Lessons Learned* (July 31, 2006).

[42] According to DOD Directive 8521.01E, §5.1.2.1 (2008), the Assistant Secretary of Defense for Research and Engineering—within the Office of the Under Secretary of Defense for Acquisition, Technology, and Logistics—is responsible for periodically assessing biometrics activities for continued effectiveness in satisfying end-user requirements.

includes a provision that would require DOD organizations to provide feedback and biometrics lessons learned to the Biometrics Identity Management Agency in its role as the Executive Manager for DOD Biometrics. In this role, the Biometrics Identity Management Agency could disseminate biometrics lessons learned collected by the various military services and combatant commands to inform relevant policies and practices. Biometrics Identity Management Agency officials told us that while they have established a process to receive Army lessons learned for biometrics, the agency does not plan to assume the additional responsibility of collecting the other military services' and combatant commands' lessons learned for biometrics issues and disseminating them across DOD without an explicit requirement to do so. By not disseminating biometrics lessons learned from existing military service and combatant command lessons learned systems across the department, DOD misses an opportunity to fully leverage its investment in biometrics.

Conclusions

U.S. military forces have used biometrics as a nonlethal weapon in counterinsurgency operations in Afghanistan to remove the anonymity sought by enemy combatants. However, issues such as minimal biometrics training for leaders; challenges with ensuring the complete, accurate, and timely transmission of biometrics data; and the absence of a requirement to disseminate biometrics lessons learned across DOD persist. As a result, these issues limit the effectiveness of biometrics as an intelligence tool and may allow enemy combatants to move more freely within and across borders.

Recommendations for Executive Action

We recommend that the Secretary of Defense take the following seven actions:

To better ensure that training supports warfighter use of biometrics, direct the military services and Special Operations Command to expand biometrics training for leaders to include

- the effective use of biometrics in combat operations,
- the importance of selecting appropriate candidates for training, and
- the importance of tracking who has completed biometrics training prior to deployment to help ensure appropriate assignments of biometrics collection responsibilities.

To better ensure the completeness and accuracy of transmitted biometrics data, direct the Assistant Secretary of Defense for Research and Engineering, through the Under Secretary of Defense for Acquisition, Technology, and Logistics, and in coordination with the military services, Special Operations Command, and Central Command, to identify and assign responsibility for biometrics data throughout the transmission process, regardless of the pathway the data travels, to include the time period between when warfighters submit their data from the biometrics collection device until the biometrics data reach ABIS.

To determine the viability and cost-effectiveness of reducing transmission times for biometrics data, direct the Assistant Secretary of Defense for Research and Engineering, through the Under Secretary of Defense for Acquisition, Technology, and Logistics, to comprehensively assess and then address, as appropriate, the factors that contribute to transmission time for biometrics data.

To more fully leverage DOD's investment in biometrics, direct the Assistant Secretary of Defense for Research and Engineering, through the Under Secretary of Defense for Acquisition, Technology, and Logistics, to

- assess the value of disseminating biometrics lessons learned from existing military service and combatant command lessons learned systems across DOD to inform relevant policies and practices, and
- implement a lessons learned dissemination process, as appropriate.

Agency Comments

We requested comments from DOD on the draft report, but none were provided. DOD did provide us with technical comments that we incorporated, as appropriate.

We are sending copies of this report to other interested congressional parties; the Secretary of Defense; the Chairman, Joint Chiefs of Staff; the Secretaries of the U.S. Army, the U.S. Navy, and the U.S. Air Force; the Commandant of the U.S. Marine Corps; and the Director, Office of Management and Budget. In addition, this report will be available at no charge on the GAO Website at http://www.gao.gov.

If you or your staff have any questions about this report, please contact me at (202) 512-4523 or at leporeb@gao.gov. Contact points for our Offices of Congressional Relations and Public Affairs may be found on the last page of this report. Key contributors to this report are listed in appendix II.

Brian J. Lepore
Director
Defense Capabilities and Management

List of Requesters

The Honorable Adam Smith
Ranking Member
Committee on Armed Services
House of Representatives

The Honorable W. "Mac" Thornberry
Chairman
The Honorable James R. Langevin
Ranking Member
Subcommittee on Emerging Threats and Capabilities
Committee on Armed Services
House of Representatives

The Honorable Jeff Miller
House of Representatives

Appendix I: Scope and Methodology

To address our audit objectives, we reviewed relevant Office of the Secretary of Defense, military service, and combatant command policies and guidance, such as the Department of Defense's (DOD) biometrics directive[1] and accompanying draft instruction,[2] and the Army's *Commander's Guide to Biometrics in Afghanistan*.[3] We obtained these and other relevant documentation, and interviewed officials from the DOD organizations identified in table 2.[4]

Table 2: DOD Organizations Visited[a]

Office of the Secretary of Defense	Office of the Under Secretary of Defense for Intelligence
	Office of the Under Secretary of Defense for Acquisition, Technology, and Logistics; Office of the Assistant Secretary of Defense, Research and Engineering; Defense Biometrics & Forensics
The Joint Staff	Force Structure, Resources, and Assessment Directorate, Requirements, Force Protection Division
U.S. Army	Headquarters, Intelligence Directorate
	Intelligence and Security Command
	Biometrics Identity Management Agency
	National Ground Intelligence Center, Charlottesville, VA
	Training and Doctrine Command Capabilities Manager – Biometrics & Forensics, Fort Huachuca, AZ
	Training and Doctrine Command, Intelligence Center of Excellence, New Systems Training and Integration Division, New Equipment Technology Team, Fort Huachuca, AZ
	Center for Army Lessons Learned, Fort Leavenworth, KS
	Project Manager for DOD Biometrics
	Project Manager for Relevant Intelligence, Surveillance, and Reconnaissance to the Edge, Aberdeen Proving Ground, MD
U.S. Navy	Office of the Assistant Secretary of the Navy for Energy, Installations, and Environment
	Office of the Deputy Assistant Secretary of the Navy for Research, Development, Test and Evaluation
	Office of Naval Intelligence

[1] DOD Directive 8521.01E, *Department of Defense Biometrics* (Feb. 21, 2008).

[2] DOD Instruction 8521.bb, *DOD Instruction on Biometrics* (draft, as of October 2011).

[3] U.S. Army, *Commander's Guide to Biometrics in Afghanistan: Observations, Insights, and Lessons*, Center for Army Lessons Learned (April 2011).

[4] We met with Navy and Air Force officials to confirm that they have collected comparatively few biometrics enrollments in Afghanistan.

Appendix I: Scope and Methodology

	Naval Criminal Investigative Service, Biometrics Division
	Office of the Chief of Naval Operations, Visit, Board, Search, and Seizure Office, Surface Warfare Directorate
U.S. Marine Corps	Headquarters, Plans, Policies, and Operations Division, Identity Operations Section
	Marine Air-Ground Task Force Integrated Systems Training Center, Camp Pendleton, CA
	Marine Forces Systems Command
	Marine Corps Center for Lessons Learned
U.S. Air Force	Headquarters, Cyberspace Operations Directorate, Cyber Surety Division
U.S. Central Command	Headquarters, Intelligence Directorate, Command, Control, Communications, Computers, Intelligence, Surveillance, and Reconnaissance Systems Branch. MacDill Air Force Base, FL
	Headquarters, Operations Directorate, MacDill Air Force Base, FL
U.S. Special Operations Command	Headquarters, Identity Operations Directorate, MacDill Air Force Base, FL
U.S. Forces-Afghanistan	Headquarters, Task Force Biometrics, Combined Joint Interagency Task Force-435, Bagram Air Field
	Biometrics Support Element – East, Bagram Air Field
	Biometrics Support Element – South, Kandahar Air Field
	Marine Corps Biometrics Liaison Office, Camp Leatherneck
	Tactical Biometrics Cell – Afghanistan, Bagram Air Field
	Tactical Biometrics Cell – Afghanistan, Kandahar Air Field
	Task Force Biometrics, Last Tactical Mile Pilot Project, Forward Operating Base Pasab
	Joint Expeditionary Forensic Facility, Kandahar Air Field
	Joint Expeditionary Forensic Facility, Camp Leatherneck
	Combined Explosives Exploitation Cell, Camp Leatherneck
	Joint Prosecution Exploitation Center, Camp Leatherneck
	Combined Joint Special Operations Task Force – Afghanistan, Camp Vance, Bagram Air Field

Source: GAO data.

Note: [a]Unless otherwise indicated, these offices and agencies are located within the Washington, D.C. metropolitan area.

To determine the extent to which DOD's biometrics training supports warfighter use of biometrics in Afghanistan, we reviewed relevant Army, Marine Corps, and Central Command policies and assessments pertaining to biometrics training for warfighters and leaders to determine the biometrics training requirements for U.S. forces operating in Afghanistan. To understand the frequency and types of biometrics training offered by the Army, Marine Corps, and Special Operations Command, we reviewed training schedules and we observed the Army's Soldier Field Service Engineer Course 2nd Pilot and Biometrics Operations Specialist/Master Gunner Training at Fort Drum, N.Y.; the

Appendix I: Scope and Methodology

Marine Corps' Biometrics Automated Toolset Basic Operator's Course at Camp Pendleton, CA; and the Special Operations Command's Technical Exploitation Course I and the Sensitive Site Exploitation Operator Advanced Courses training at Fort Bragg, N.C.. In addition, we discussed training with military officials in Afghanistan from the organizations listed in table 2.

To determine the extent to which DOD is effectively collecting and transmitting biometrics data, we obtained, reviewed, and analyzed relevant Central Command issued Joint Urgent Operational Need Statements. In addition, we reviewed documents on biometrics collections and transmissions and spoke with Office of the Secretary of Defense, Army, Marine Corps, Central Command, and Special Operations Command officials. We reviewed DOD biometrics submission latency data to understand data transmission over time. We assessed the reliability of the data by reviewing related documentation and interviewing knowledgeable officials. Although we found the data sufficiently reliable to provide descriptive and summary statistics, problems were identified with the completeness and accuracy of the data due to external factors, such as inaccurate time/date stamps on biometrics collection devices. As a result, we developed a recommendation to assign responsibility for biometrics data throughout the transmission process, to include the time period between when warfighters submit their data into the Biometrics Automated Toolset system until the biometrics data reach ABIS to better ensure completeness and accuracy of biometrics data during the transmission process. We also reviewed the *Standards for Internal Control in the Federal Government* for information on data completeness and accuracy assurance.[5] We conducted site visits to four military installations in Afghanistan to ascertain how biometrics are being collected, utilized, and transmitted. Specifically, we visited Bagram Air Field, Kandahar Air Field, Marine Corps Base Camp Leatherneck, and Forward Operating Base Pasab to meet with military officials responsible for leading and performing biometrics collection, analysis, and transmission activities in Afghanistan and for operating the Last Tactical Mile pilot project.

---

[5] GAO, *Standards for Internal Control in the Federal Government,* GAO/AIMD-00-21.3.1 (Washington, D.C.: November 1999).

Appendix I: Scope and Methodology

To determine the extent to which DOD has developed a process to collect and disseminate biometrics lessons learned, we analyzed relevant Office of the Secretary of Defense, Army, and Marine Corps guidance and policies, and met with officials from each of these organizations to discuss current practices.

We conducted this performance audit from May 2011 through April 2012 in accordance with generally accepted government auditing standards. Those standards require that we plan and perform the audit to obtain sufficient and appropriate evidence to provide a reasonable basis for our findings and conclusions based on our audit objectives. We believe that the evidence obtained provides a reasonable basis for our findings and conclusions based on our audit objectives.

Appendix II: GAO Contact and Staff Acknowledgments

GAO Contact	Brian J. Lepore, Director, 202-512-4523 or leporeb@gao.gov
Acknowledgments	In addition to the contact named above, Marc Schwartz, Assistant Director; Grace Coleman; Mary Coyle; Davi M. D'Agostino, Director (retired); Bethann E. Ritter; Amie Steele; and Spencer Tacktill made key contributions to this report. Ashley Alley, Timothy Persons, Terry Richardson, John Van Schaik, and Michael Willems provided technical assistance.

Related GAO Products

Defense Biometrics: DOD Can Better Conform to Standards and Share Biometric Information with Federal Agencies. GAO-11-276. Washington, D.C.: March 31, 2011.

Defense Management: DOD Can Establish More Guidance for Biometrics Collection and Explore Broader Data Sharing. GAO-09-49. Washington, D.C.: October 15, 2008.

Defense Management: DOD Needs to Establish Clear Goals and Objectives, Guidance, and a Designated Budget to Manage its Biometrics Activities. GAO-08-1065. Washington, D.C.: September 26, 2008.

GAO's Mission	The Government Accountability Office, the audit, evaluation, and investigative arm of Congress, exists to support Congress in meeting its constitutional responsibilities and to help improve the performance and accountability of the federal government for the American people. GAO examines the use of public funds; evaluates federal programs and policies; and provides analyses, recommendations, and other assistance to help Congress make informed oversight, policy, and funding decisions. GAO's commitment to good government is reflected in its core values of accountability, integrity, and reliability.
Obtaining Copies of GAO Reports and Testimony	The fastest and easiest way to obtain copies of GAO documents at no cost is through GAO's Website (www.gao.gov). Each weekday afternoon, GAO posts on its Website newly released reports, testimony, and correspondence. To have GAO e-mail you a list of newly posted products, go to www.gao.gov and select "E-mail Updates."
Order by Phone	The price of each GAO publication reflects GAO's actual cost of production and distribution and depends on the number of pages in the publication and whether the publication is printed in color or black and white. Pricing and ordering information is posted on GAO's Website, http://www.gao.gov/ordering.htm. Place orders by calling (202) 512-6000, toll free (866) 801-7077, or TDD (202) 512-2537. Orders may be paid for using American Express, Discover Card, MasterCard, Visa, check, or money order. Call for additional information.
Connect with GAO	Connect with GAO on Facebook, Flickr, Twitter, and YouTube. Subscribe to our RSS Feeds or E-mail Updates. Listen to our Podcasts. Visit GAO on the Web at www.gao.gov.
To Report Fraud, Waste, and Abuse in Federal Programs	Contact: Website: www.gao.gov/fraudnet/fraudnet.htm E-mail: fraudnet@gao.gov Automated answering system: (800) 424-5454 or (202) 512-7470
Congressional Relations	Katherine Siggerud, Managing Director, siggerudk@gao.gov, (202) 512-4400, U.S. Government Accountability Office, 441 G Street NW, Room 7125, Washington, DC 20548
Public Affairs	Chuck Young, Managing Director, youngc1@gao.gov, (202) 512-4800 U.S. Government Accountability Office, 441 G Street NW, Room 7149 Washington, DC 20548

Please Print on Recycled Paper.

www.ingramcontent.com/pod-product-compliance
Lightning Source LLC
Chambersburg PA
CBHW081803170526
45167CB00008B/3305